Alexander R Becker

Gun-Shot Wounds

Particularly those Caused by Newly-Invented Missiles

Alexander R Becker

Gun-Shot Wounds
Particularly those Caused by Newly-Invented Missiles

ISBN/EAN: 9783337184421

Printed in Europe, USA, Canada, Australia, Japan

Cover: Foto ©berggeist007 / pixelio.de

More available books at **www.hansebooks.com**

GUN-SHOT WOUNDS,

PARTICULARLY THOSE CAUSED BY NEWLY-INVENTED MISSILES.

AN ESSAY WHICH RECEIVED THE FISKE FUND PREMIUM OF THE RHODE ISLAND
MEDICAL SOCIETY, FOR 1864.

BY ALEXANDER R. BECKER., M.D., PROVIDENCE, R. I.

" Arma virumque cano . . . . "

[Reprinted from the Boston Medical and Surgical Journal, February, 1865.]

THE subject of " Gun-shot Wounds " is one of great interest, especially at the present time, when our country is engaged in a war which calls forth such large numbers of her sons, and exposes them to all the hardships and injuries of the field.

To treat this subject as fully as it deserves is, of course, impossible in an essay of this sort. I have therefore endeavored in the following pages to give a somewhat extended summary of the injuries which occur on the battle-field, with their consequences and treatment, as given by the best authorities, aided by my comparatively slight experience in the Army of the Potomac during the Peninsular Campaign.

For statistics I have been obliged to refer to Dr. Macleod's valuable work on the "Surgery of the Crimean War," as we have none, as yet, of the present war which are sufficiently connected to be of any value.

CHAPTER I.

The Peculiarities of Gun-shot Wounds, and their general Treatment.

The sensation caused by a gun-shot wound in a fleshy part, is usually described by the sufferer as resembling the effect of a smart blow from a supple cane. Some, however, feel as if a red-hot wire were passed through the part; while others are entirely unconscious of any wound, and are first apprised of it by the flowing of blood. This seems impossible; but Macleod, in his " Notes on the Surgery of the Crimean War," affirms that he personally " knew an officer who had both legs carried away, and who said it was only when he attempted to rise that he became aware of the injury received." The present war has also afforded instances of a similar character.

The " collapse " and mental trepidation which frequently follow the infliction of a wound, though useful as a diagnostic indication of the gravity of the injury, is not entirely to be relied on; for this often depends as much on the frame and "nerve" of the patient as on the severity of the wound. It is, however, important to remember that the "shock" succeeding gun-shot wounds is greater when the lower extremities are injured than when the arms suffer; and Chevalier makes the remark, that the shock is always greater when a ball strikes a muscle in action than when it impinges against one which is relaxed.

The destruction caused by a ball depends on its shape and velocity, the distance at which it is fired, the direction of its flight, and on the part struck. If fragments of metal are fired—as sometimes happens during sieges and riots—a very lacerated, irregular and dangerous wound may be produced. The shape, great velocity and peculiar motion of the conical ball gives a very different character to its wounds from that caused by the round musket-ball. If, when fired at short range, it strike a fleshy part, the conical ball produces, I think, less laceration than the old ball. But if, when fired at long range, it strike a bony part, with but little covering of flesh, such as the hand or foot, then the tearing, especially at the place of exit, is much more marked. The cause of this will be found in its retarded velocity, and also in the flattening of the ball against the bone, its diameter being thus increased before it escapes.

It is not always so easy, as the description of authors would lead one to suppose, to distinguish the wound of entrance from that of exit. That the former is more regular than the latter is generally true; but that the lips of the one are inverted while those of the other are everted, is seldom clearly evident. If the speed of the ball be great, and no bone struck, the size and discoloration of the wounds is very nearly the same. But if its flight be so far spent as to be retarded by contact with the body, especially if it encounter a bone or strong aponeurosis, then the wound of exit will be considerably larger than that of entrance. And this is particularly true of wounds caused by conical balls.

In M. Arnel's experiments, given in the *Journal Univer. de Médecine* for 1830, it is shown that a ball fired against a number of boards firmly bound together causes a series of holes, progressively increasing in size, so that a cone is formed by their union, whose base is represented by the last exit hole. M. Devergie's experiments on the same point also go to prove this.

To the military surgeon, and also to the medico-legal jurist, it is often of great importance to be able to determine whether the two apertures in a patient's limb have been caused by one ball, which is thus known to have escaped, or by two balls still remaining in the limb. The following instance, related by an Indian surgeon, will serve to illustrate this. A wound was found below and another above the patella of a soldier. The former had all the appearance of the wound of entrance, and the latter of the wound of exit. The opening of an abscess in the thigh, a fortnight after, gave exit to a grape-shot, and it was found that each opening had been caused by a different ball.

M. Begin has made the following valuable observation in regard to the resulting cicatrices. That of entrance, he says, is generally white, depressed, and often adherent to the underlying parts; while that of exit is only a sort of irregular spot, which does not adhere to the parts below, and is sometimes so indistinct as to be concealed in the folds of the skin.

One marked characteristic of gun-shot wounds is, their healing only by second intention. Occasionally, however, an exception to this rule occurs. Dr. Stewart, of the English Army, reports a case which occurred during the Caffre war. A Fingo received a wound in the muscles of the back, and union without suppuration took place. Two things, however, must be necessary to produce so happy a result—a most healthy and temperate patient, and a very rapid flight of the ball.

Some very extraordinary cases are recorded, where a large body has entered through muscle, leaving so little appearance of its presence as entirely to escape notice. Baron Larrey mentions a case in which a ball weighing *five* pounds was extracted by him from the thigh of a soldier. The presence of so large a body had not been detected by the surgeon in charge, and the patient experienced no inconvenience from it, except a feeling of weight in the limb. Hennen, too, mentions a case which occurred at the battle of Seringapatam, in which a *twelve*-pound shot was found in the thigh of an officer; and "so little appearance was there of a body of such bulk that he was brought to the camp, where he soon expired, without any suspicion of the presence of the ball till it was discovered on examination."

Baudens affirms, that when a ball is cut out from among muscles, it is enclosed in a cellular envelope, which he calls "kyste primatif," as contrasted with the "kyste definitif," which forms its sac when it has long remained within the tissues. I have never seen this confirmed by any other author, nor have I been able to discover it ·myself.

Unless special precautions are taken to prevent the contraction of muscles which have been severely injured by ball, this most disagreeable result will be very apt to occur.

Tendons, from their toughness, elasticity, form and mobility, will often escape with little or no injury, especially if they are relaxed when struck. A round ball, especially if its force be nearly expended, and it should strike at an angle to the surface, is often deflected from its course by a strong aponeurosis like the "fascia lata." A conical ball is seldom, if ever, so turned.

But it is on bone that we see the most destructive effects of a ball. In regard to the old ball:—1st. If its line of flight is very oblique, and it strike against a flat bone, it may be thrown off, causing no other damage than depriving the bone of its periosteum. But if this occur with the bones of the head, great danger may ensue, as will be shown in another place. And such wounds of the long bones, though apparently trifling at first, are frequently very serious in their results; both from subsequent disease of the bone, on account of the detachment of the periosteum, and from inflammation being set up in the medullary canal. 2d. It may sometimes be flattened against the shaft of a long bone, without subsequent injury. This, however, can only occur when the ball is almost "spent." 3d. It may turn round a bone without injuring it. 4th. It may notch or partly perforate a long bone; may even enter the medullary cavity and remain there, causing, of course, the most dangerous symptoms. 5th. If the force be a little greater, it may split the bone longitudinally, without causing a transverse fracture. A case of this sort is related by Leveillé, and quoted by Malgaigne, in which an Austrian soldier, at Marengo, was struck by a ball in the lower third of the leg. He walked several miles to the rear, where he was seen, and his wound thought to be very slight. A superficial exfoliation of the bone was alone expected. His symptoms, however, became so serious that the leg was removed, when it was found that from the place where the impression of the ball existed, there proceeded several longitudal and oblique clefts, which extended from the lower third of the tibia up to near the head of the bone. 6th. Into the spongy heads of bones a ball may be driven as into a plank of wood, with little or no splintering. 7th. It may pass through, causing a clean hole.

The conical ball, however, never acts in any of these ways. It is seldom, if ever, split itself, and always splinters the bone against which it strikes, to a greater or less extent, and that in the direction of the bone's axis. This tendency to splitting shows itself much more in a downward, than in an upward direction; so that the destruction which this ball occasions will be greater when it strikes the upper than the lower end of the shaft.

As a general thing any ball, no matter what its shape, will fracture and split a bone, if it strike about the middle of the shaft. But while the round ball causes but little comminution, the conical ball, especially if it have a broad deep cup at its base, like the Enfield rifle-ball, splits and fractures the bone to such an extent that large spiculæ are detached, and smaller ones are driven in all directions into the neighboring parts; and this is what renders fractures of the long bones, by the new ball, so dangerous.

There is no doubt but that a ball may remain imbedded in bone for a life-time, with very little if any inconvenience resulting therefrom; but this is decidedly the exception. Guthrie is very emphatic

in his directions to remove balls so placed, and predicts the most
unfortunate results if this be neglected. And Malgaigne, after nar-
rating several instances in which balls have remained without causing harm, concludes thus:—" It is necessary to mention these fortunate cases, as evidence of the resources of nature; but they hardly
serve to weaken the force of the prognosis, when a ball cannot be
extracted, or the essential indication of this sort of lesion, the extraction of a foreign body. This indication is, then, of the first
importance."

Nerves usually escape injury by a ball. Still, if the ball has been
rendered jagged by previous contact with some hard substance, it
may cause grave injury even to the largest nerves, especially when
they lie in close proximity to a bone. The paralysis which succeeds
an injury to a nerve may come on at once, or be for a long time de-
layed; and may or may not be accompanied by pain. For instance,
the hand has several times been known to waste, without pain, when
some of its nerves have been injured.

But making all due allowance for the form, strong coat, and elasticity of arteries, it is wonderful how frequently they escape injury
from gun-shot wounds. The rarity of primary hæmorrhage on the
field is well known; and yet we constantly see wounds through the
very course of large arteries. When primary hæmorrhage does occur on the field, it is more frequently from the veins, which are more
easily cut than arteries.

We have fewer instances of the wandering of balls now than for-
merly. The conical ball generally takes the shortest route to and
through its mark, and sometimes even further; for cases have been
related where one ball has been known to pass through the bodies of
two men and lodge in a third. But of the round ball many curious
stories are told. Dr. Macleod, to whom I have before referred, nar-
rates two instances: in one of these, the ball entered above the elbow and was removed from the opposite maxilla; and in the other
it entered the right hip and was removed from the left popliteal
space.

In examining a gun-shot wound, we must always bear in mind the
possibility of some foreign substance, as a piece of the man's clothing or accoutrements, having entered with the ball; and its presence
in the wound is often far more troublesome than the ball itself.

There has been a great deal of discussion and diversity of opinion in regard to the extraction of balls. Begin's precept, as given
to the Academy, seems to me to coincide with both science and common sense. " Selon moi," he says, " l'indication de leur extraction
est toujours presente; toujours le chirurgeon doit chercher à la rem-
plir; mais il doit le faire avec la prudence, et la measure que la rai-
son conseille. S'il reussit, il aura beaucoup fait en faveur du blessé.
S'il s'arrête devant l'impossibilité absolue, ou devant la crainte de
produire les lesions additionelles trop grave, il aura encore satisfait

aux principes de l'art; et quel que soient les resultats de la blessure, il n'aura pas a se reprocher de les avoir laisse devenir funestes par son inertie." If we compare the opinions of surgeons, we find that while civilians frequently consider it a matter of secondary importance, military surgeons always place great weight upon its accomplishment. The question is, not whether balls *may* remain in the body without causing harm, but whether· they do so in the majority of cases. Statistics prove conclusively that in the majority of cases they *do* cause harm and annoyance. Consequently, as "science is not made up of exceptions," and it is the unquestioned duty of the medical man, in treating his patient, to consider his permanent welfare rather than his temporary comfort, it appears to me that the early extraction of the ball in wounds of this sort should *always* be accomplished when practicable.

We have in this country had very few opportunities of learning the subsequent history of men who carried unextracted balls. But M. Hutin, chief surgeon of the "Hôtel des Invalides," of France, tells us that while 4,000 cases had been examined by him in five years, only 12 men presented themselves who suffered no inconvenience from unextracted balls, while the wounds of 200 continued to open and close until the foreign body had been extracted. This evidence, coming from such a source, is, to my mind, conclusive. For there is no other institution in the world, to my knowledge, where the opportunities for examining this class of cases are so complete.

The extraction of a ball is not generally difficult, particularly if we can see the patient soon after the infliction of his wound. It is of great importance to extract the ball, if possible, before inflammation has set in; as otherwise the wound is more or less closed, we cause more pain, and our chance of easily finding the ball is diminished.

The first thing is, to place the patient in the same position, as nearly as possible, as that which he occupied at the time of injury; also to place ourselves, relatively to him, in the direction from which the ball came. Taking into consideration what effect the bones and tendons in the neighborhood may have had in deflecting the ball, and consulting the patient's sensations, we shall generally, without much difficulty, succeed in finding it. By examining the clothing, we shall discover whether any part of it has been carried into the wound; and we should remember that sometimes a ball will carry in a cul-de-sac of the clothes and be withdrawn by it. I have before alluded to the awkward mistakes which sometimes occur when two wounds, having the appearance of the wounds of entrance and exit, are in reality caused by two balls. This should always be borne in mind when making such an examination.

Macleod, to show the importance of an early and careful examination, as well as that we should never rely too much on the patient's statement, relates the following case as having happened to himself in the Crimea. " A soldier wounded on the 18th of June came under my care in the general hospital. His right arm, which had been fractured compoundly, was greatly swollen at the time of admission. I was told, and accepted the story, that the accident had been caused by a piece of shell, to which species of injury the wound bore every resemblance, and that it had been removed in the trenches. At the earnest solicitation of the patient, I contented myself with applying the apparatus for saving the limb, without minutely examining the wound. When removing the limb at the shoulder, a few days afterwards, to my great astonishment a large grape-shot dropped from among the muscles."

Sir Charles Bell has shown how the nerves may indicate to us the course, and sometimes the position of the ball. He says, "So when a ball has taken its course through the pelvis, or across the shoulder, the defect of feeling in the extremity, being studied anatomically, will inform you of its course; that it has cut, or is pressing on a certain trunk of nerves."

We should always make certain of the position of a ball, immediately before taking any steps for its removal; remembering the rule laid down by Dupuytren, never to act upon information regarding the site of a ball, obtained the day before, from the very rapid manner in which they often shift from one spot to another.

If the wound be large, as it generally is from a conical ball, the finger is the best probe; otherwise, or if the wound be too deep for the finger, a large gum-elastic bougie is the best substitute.

There have been an infinite variety of instruments invented for the extraction of balls. Macleod prefers the common dressing forceps, if long enough and fine enough in the handle. Larrey used the polypus forceps in preference to anything else; and I must say that I agree with him in thinking them the most useful and convenient.

It is of great importance to sustain the limb with the disengaged hand, on the side opposite to that at which we introduce the forceps. If the ball lie at all near the surface, and especially if its course has been from above downwards, we should always cut upon it; as by this course we facilitate its removal, and provide an opening for the pus. The long continuance of the discharge, its gleety character, and the persistence of pain in the track, is almost always occasioned by the presence of some foreign body—it may be a mere shred—in the wound.

The constitutional fever which succeeds a gun-shot wound is generally, though not always, in proportion to the part injured. The fever will frequently be of an endemic or epidemic character; but in war the tendency seems chiefly to be to a low typhoid type.

The mitigation of the constitutional fever and of the local inflammation; the prevention of all accumulations of pus, by making judicious escapes for it; the application of light, unirritating dressings; rest, and attention to the essential principles of all surgery, comprise the general treatment which gun-shot wounds usually require.

In the early stages cold, even ice as recommended by Baudens, may be of great use; and in wounds of the hand and fore-arm, irrigation is said to be of service. But when inflammation and suppuration are present, hot applications will almost always be found the most beneficial, as well as the most grateful to the patient. Strict attention to the position of the limb is of the utmost consequence.

Soldiers in war are easily depressed, and should not be too sparingly fed, when admitted to hospital, unless suffering from a wound of the head, chest or abdomen. There is too much tendency to look upon gun-shot wounds, as a class, as highly inflammatory, and to treat them accordingly. Velpeau's rule, and it seems to me to be a common-sense one, is to remove his wounded and operated on as little as possible from their ordinary diet, when hungry, and when there is no disturbance of the digestive or circulatory systems.

By gentle syringing with lukewarm water, from one opening to the other, we get rid of any shreds of cloth, clots of blood, pus, &c., which may be lodged in the wound, sustaining the suppuration, with very little disturbance of the parts. The French are in the habit of employing, with apparent advantage, a lotion composed of one part of the perchloride of iron and three parts of water, in profusely suppurating wounds.

Shell wounds and grazes by round shot are often attended with much injury, deep seated, and often little suspected; and not unfrequently result in wide-spread sloughing of the soft parts. The following case, which occurred at Sadoolapore, is an excellent illustration of this, and is an instance of what in former times would have been set down to the wind of the ball. " Private Conally was hit by a round shot on the outer side of the right arm and thorax. A blue mark alone was occasioned on the arm, and little or no mark was found on the chest. He died in twenty hours, without having rallied from the shock. The peritoneal cavity was found full of dark blood. The right lobe of the liver was torn into small pieces, some of which were loose and mixed with blood. There was no sign of inflammation of the peritoneum, and the other viscera were healthy."

There have been many instances of the very near approach, and even slight contact of round shot, without any further inconvenience arising therefrom than might naturally be looked for from the unexpected and unwelcome vicinity of such an intruder.

CHAPTER II.

Chloroform ; Primary and Secondary Hæmorrhage ; Tetanus ; Gangrene.

NOWHERE do we so thoroughly appreciate the advantages of anæsthetics as in the field. We are enabled by them to soothe the agony produced by the terrible wounds which continually come under our notice in war, and also with their aid to perform many operations which were otherwise impossible. Macleod gives his opinion very decidedly on this subject in these words:—"For my own part, I never had reason for one moment to doubt the unfailing good and universal applicability of chloroform in gun-shot wounds, *if properly administered*. I most conscientiously believe that its use in our army directly saved many lives; that many operations necessary for this end were performed by its assistance which could not otherwise have been attempted; that these operations were more successfully, because more carefully executed; that life was often saved even by the avoidance of pain; the *morale* of the wounded better sustained, and the courage and comfort of the surgeon increased. . . . . . . I think I have seen enough of its effects to conclude that if its action is not carried beyond the stage necessary for operation, it does not increase the depression which results from injury, but that, on the contrary, it in many instances supports the strength under operation." Many prefer ether to chloroform, believing the latter to be a most dangerous agent. But if the surgeon will assure himself that there exists no disease of the heart, brain or lungs in the patient; if proper care be taken to give free access to the atmospheric air, and the pulse of the patient be carefully watched, I do not believe that any pernicious effects will be found to follow the use of chloroform. (To men who have lost much blood, however, it must, of course be administered with great care, from the rapidity of its absorption by such persons.) Our distinguished surgeon, J. M. Carnochan, M.D.P., of New York, has been kind enough to allow me to quote his experience on this subject. He says:—"I have administered chloroform in about three thousand (3,000) cases, and so far I have never had any fatal effects resulting from its administration in my hands. . . . . . . I would, moreover, state that I invariably use the pure Scotch chloroform, regarding ether, or the combination of ether and chloroform, as recommended by some physicians, as entirely inferior."

Chloroform is more applicable to field service : 1st, because a much less quantity is needful to bring a patient under its anæsthetic influence; 2d, because its action is more prompt; and, 3d, because it is much less likely to produce nausea and vomiting than ether. And these are important considerations, where time is so valuable.

Hæmorrhage was in old times the Military Surgeon's great bugbear. Since, however, we have found out that it is not of such frequent occurrence on the field as was supposed, and the means of arresting it are better understood, it is not so much feared. There is, it is true, a gush of blood when an artery is opened by a ball; but the artery soon retracts and closes itself. Larrey mentions a case where a soldier, struck on the lower third of the thigh by a ball, suffered one severe hæmorrhage, which was never repeated. The limb became cold, the popliteal ceased to beat, and the ends of the divided femoral could be felt when the finger was placed in the wound. This man recovered perfectly. And the younger Larrey records a case from the wounded of the siege of Antwerp. A shell passed between a man's thighs, and, destroying the soft parts, divided both femorals; yet there was no hæmorrhage, although the pulsation continued in the upper ends of the vessels to within a few lines of their extremities.

There has been a great deal said about the advantage of carrying a number of tourniquets into the field, to arrest primary hæmorrhage; but the surgeon's place in battle being at the rear, he has but little opportunity to carry this suggestion into effect. An intelligent soldier, however, can easily make a tourniquet with his handkerchief and bayonet; or a piece of stick.

I have before alluded to the strange manner in which arteries sometimes escape injury. The following case came under my notice during the Peninsular Campaign in Virginia. A ball passed through the neck of a soldier, just posterior to the carotids on both sides, and anterior to the spine. This man recovered perfectly in a very short time, not having had a bad symptom. Dr. Macleod relates two or three cases, so remarkable that I cannot help quoting them. "A soldier of the Buffs was wounded by a rifle-ball, which struck him in the nape of the neck. It passed forwards round the right side of the neck, going deeply through the tissues; turning up under the angle of the inferior maxilla, it fractured the superior maxillary and malar bones, destroyed the eye, and escaped, killing a man who was sitting beside him. This patient made a rapid recovery.—A French soldier, at the Alma, was struck obliquely by a rifle-ball, near to but outside the right nipple; the ball passed seemingly quite through the vessels and nerves in the axilla, and escaped behind. His cure was rapid and uninterrupted.—Another Frenchman was struck in the trenches by a ball, a little below the middle of the right clavicle. The ball escaped behind, breaking off the upper third of the posterior border of the scapula, and yet he recovered perfectly, without any bleeding taking place."

I had, last June, the opportunity of seeing a very interesting case at Hampton Hospital, Virginia, which was under the care of my friend Dr. Charles F. Bullen, of Montreal, C. E. (at that time Act. Assis't Surgeon U.S.A.), and of which he was kind enough to give me a copy of his notes. "Adam Grimm, private, Co. D, 7th Conn. Vols., aged 21, was wounded before Petersburg June 9th, 1864, by a rifle-ball, which fractured the acromion end of the right clavicle, passed beneath the scapula and out below the lower border. On admission to the Hospital, three days after the injury, some fragments of bone were removed. The wound looked healthy, and continued discharging laudable pus and granulating till June 28th, 11, A.M., when secondary hæmorrhage occurred. He then lost about six ounces of blood before it was checked by pressure.

June 29th, 10, A.M.—Hæmorrhage again occurred, more severely than before, losing from fourteen to sixteen ounces of blood. The cavity of the wound was by this time much enlarged. The hæmorrhage was again apparently checked by plugging the wound with lint saturated with perchloride of iron. But in two hours the whole of the tissues between the wound and the neck were engorged with blood, the swelling rapidly increasing. and thus showing that he was still bleeding. After consultation, it was decided to stimulate freely and give narcotics to relieve pain, and let him remain till morning.

June 30th, 11, A.M., being in about the same condition—the tongue dry and glazed, pulse 120 and very weak, and with the engorgement gradually increasing—the subclavian was ligated successfully in the first part of its course. Coagula were then removed from the cavity of the wound, and it was syringed out with ice-water, no bleeding being apparent. Immediately after the operation he rallied; the tongue became moist; pulse at left wrist 110, at right wrist *none*. The temperature of both arms was the same, and continued so throughout.

July 1st, 10, A.M.—Left pulse 110, right barely perceptible. Patient in good spirits; takes nourishment freely, but complains of pain in swallowing. 10, P.M.—Left pulse 112, right same as in the morning. Ordered ℞. Liq. ammon. acetat., ℥ i.; tinct. aconit. ♏v.; to be taken every four hours.

2d.—Left pulse 110, right increasing a little in strength; no pain in swallowing, and improving.

3d.—Left pulse 108, right same as yesterday.

4th.—Left pulse 100, right same as yesterday; takes nourishment freely, and both wounds looking healthy and well.

5th.—Left pulse 96, right same as yesterday.

6th.—Left pulse 90, right same as yesterday. Omit medicine.

7th.—Left pulse 90, right same as yesterday. Complains of pain in the region of the heart, but no abnormal sounds heard.

8th.—Left pulse 120, right same as before; tongue dry and glazed. At 9, P.M., he had a rigor.

9th, 7, A.M.—A slight hæmorrhage from the point where the artery was ligated. The wound was plugged and pressure employed. At 10, A.M., the hæmorrhage recurred more severely than before. From this time until evening there were repeated hæmorrhages; the patient gradually sank, and died at 8, P.M., remaining sensible to the last.

Autopsy.—Both the superscapular and posterior scapular arteries were found to be in a sloughing condition, which was apparently the cause of the last hæmorrhages. The subclavian was ligated about half an inch from its origin. The ligature had come away, and the coats of the artery were ulcerated through. On the cardiac side a slight clot had formed, but on the distal side the clot was larger, firmer, and more perfectly organized."

This case is exceedingly interesting, both on account of the infrequency of the operation and because the man lived so long after its performance—nine days and eight hours; and at one time it really seemed as if he would recover.

Hæmorrhage should be divided into three periods: "primary," occurring within twenty-four hours; "intermediary," between that and the tenth day, caused by sloughing resulting directly from the injury; and "secondary," that which takes place at a later date, from ulceration or other morbid action. There have been various opinions in regard to the period at which hæmorrhage is most likely to occur. Guthrie says from the eighth to the twentieth day; Dupuytren, from the tenth to the twentieth; Henman, from the fifth to the eleventh; Roux, from the sixth to the twentieth; and Macleod, from the fifth to the twenty-fifth—the majority appearing on the fifteenth day after the receipt of the wound.

In field practice, it is often impossible to give the wounded the rest and quiet of mind and body which they require, and by which secondary hæmorrhage would frequently be avoided. Previous debility will also be likely to favor hæmorrhage; as has occurred in many cases in our army, when the men were more or less exhausted by continued diarrhœa.

Hæmorrhage should be controlled by pressure, cold, styptics, ligature of the end of the artery, and, in severe cases, by ligature of the main trunk of the artery.

Tetanus, fortunately, is a very rare complication. Alcock sets the proportion at one in seventy-nine; but in the Crimea, Macleod was only able to discover thirteen cases. I have not been able to find sufficient data to enable me to fix the frequency of its occurrence in our army. Opinions vary in regard to the causes of this dreadful disease; but it is certain that unextracted balls are a prolific cause of its development. In the Indies, heat is considered as the cause; while Dr. Kane gives intense cold as the cause of the great number of cases which occurred among his men.

Romberg sums up his view of the treatment thus:—"The results of treatment amount to this, that whenever tetanus puts on the acute form, no curative proceeding will prevail; whilst in the milder and more tardy forms, the most various remedies have been followed by a cure." Larrey trusted most to opium and camphor, with section of the nerve in cases adapted for it. Opium and chloroform seem, however, to have the greatest amount of evidence in their favor.

I will quote one instance of tetanus from Dr. Macleod's excellent work. "Barker, a private of the 38th Reg't, aged 20, was admitted into the general hospital in camp, June 18th. A ball had penetrated his left thigh at its inner and lower aspect, and lodged. Four days afterwards it was found near the wound, and removed. By the 28th, the wound was looking sloughy, and the discharge was thin and unhealthy. He complained much more about his wound than was usual, and appeared very anxious. On the 30th, I noticed some twitching of the limb as it was being dressed. His bowels were free, but he complained of sleeping little at night. The wound was freely enlarged, and covered with a poultice. He was purged with croton oil and clysters. He grew gradually worse. During the two succeeding days the spasms were very decidedly pronounced over the left side. He described them himself as proceeding in 'flashes' from his wound to the spine, and back again. Touching the limb, and especially the sole of the foot, immediately aroused the most violent spasmodic contractions. His pulse rose to 92, and his respirations to 29 per minute. He did not complain of pain, but was greatly distressed by a thick spit, which clung to his teeth, and which he was always making violent efforts to expel. The left side of the body was almost alone affected, and the spasms drew him diagonally backwards, and to the wounded side. He had no trismus for the first day, but afterwards it became marked. He always said, that he was sure if he could only sleep he would be all right. I brought him under the influence of chloroform, and while its effects continued the spasms were relieved, and certainly the pulse and respirations were reduced in frequency; but as soon as he awoke, all his worst symptoms returned in undiminished vigor. Having seen the utter futility of chloroform to relieve the spasms permanently, or to arrest the disease, in two former cases at home, where the anæsthetic had been fairly tried, I determined to abandon it and trust to opium. This, with enemas, nourishing food, and local emollient applications, comprehended all the treatment. The symptoms were not abated, except for short intervals, and then only in proportion as sleep was procured. His skin was always covered with an odorous perspiration. The abdomen got distended and hard. The muscles of the back were markedly hard and contracted, particularly on the left side. The left leg was stretched out spasmodically, every muscle defined. The right limb was drawn up, and he lay across the bed. The wound was sloughy, and shreds of fascia escaped with the discharge. The urine became scanty and high colored, and required to be drawn off by the catheter. Eventually he suffered much pain in the left groin and calf of the leg, as well as at the ensiform cartilage. When trying to raise himself on his elbow, on the fifth day of the attack, and seventeenth after admission, he was violently convulsed, so that he was bent greatly backwards; he put his hand to his throat as if choking, and fell back dead. The wound was found to be lined with an ashy slough. The bone was not injured. The fascia lata was much torn, and was pierced and ulcerated at a spot on the anterior and external aspect of the limb, some little distance from the wound. The ball had evidently penetrated to this point. No nerve fibres could be detected near the wound. The parts in the neighborhood were sound. The brain and internal organs were healthy. The lungs were only slightly congested, and viscid mucus was present in the larger tubes. The spinal canal contained a good deal of fluid blood. The cord and its membranes were congested. In the lower cervical and upper dorsal regions, the substance of the cord was varicose—contracted and expanded into a series of knots. There was no other pathological appearance."

I should add, that Sir James M'Gregor affirms that this disease never makes its appearance after the twenty-second day.

We now come to another fearful complication of gun-shot wounds —*Gangrene*. It is sometimes contagious, and spreads over whole wards and hospitals; but more frequently, I think, it attacks only those whose general standard of health is greatly lowered. The earliest symptom is pain in the part, which sometimes precedes the ulcerative process by a couple of days. As a general thing, the edges of the wound do not swell up, but remain thin while they are undermined. The pain usually continues during the process of destruction. It appears chiefly in the lower extremities, and in wounds whose progress towards cure has been for some time stationary. It seldom burrows far into the intermuscular tissues, but confines its ravages to the surface and circumference of the wound.

In the treatment of gangrene there are two things to be accomplished: 1st, to arrest the sloughing process; and 2d, to keep up the general condition of the patient by every possible means. For a local application, the permanganate of potash, or, still better, the concentrated muriatic or nitric acids, will be found the most useful. These applications should be followed by anodyne poultices, as yeast and lupulin—the lupulin for its anodyne properties, and the yeast to aid in the separation of the slough by fermentation. Opium should also be administered freely (internally)—six to eight or ten grains per day. Opium is useful, and in fact necessary, to produce sleep and to allay the pain and nausea. It is also thought to aid in restoring the healthy action through the capillaries. The constitutional treatment should be continued by the administration of tinct. ferri, $\mathfrak{m}$xx., or even $\mathfrak{m}$xxx., three times daily. Porter is also of much benefit, and two or three pint bottles may be given daily. And milk punch, or, still better, eggnog is almost invaluable; it should be made in the proportion of two thirds milk and one third brandy. In short, the local treatment should be of the strongest destructive character, and the constitutional treatment of the strongest tonic character. There are undoubtedly two kinds of gangrene, but I am inclined to think that they depend entirely upon the constitution of the patient and upon the influences to which ho has previously been exposed. There is one highly predisposing cause of gangrene which I do not think is properly appreciated by the greater number of military surgeons, and that is the close proximity of fever and chronic diarrhœa patients to wounded men. I know that it is frequently impossible in military hospitals, where they are crowded for room, and with new cases constantly arriving, to classify wounds and different diseases and keep them separate; but I think it might be done far more than it is, and I am satisfied that the admixture of fever and diarrhœa cases with wounded men exercises a most deleterious influence upon the latter; and predisposes their wounds to take on unhealthy action. And I also believe that if these cases were kept separate there would be far less gangrene in our military hospitals.

The notes of the following cases were kindly furnished me by my friend Dr. D. R. Brower, Ass't Surg. U.S.V., who was in charge of the Gangrene Camp at Hampton Hospital, Virginia, from May 22d to June 29th, 1864. During that time he treated one hundred and one (101) cases of gangrene without a single death, and this is a record which I think few men can equal.

Thos. B. Benedict, aged 32, private, Co. D, 7th Conn. Vols., received, on May 10th, a flesh wound of the right thigh from a Minié ball. The wound did excellently well until May 23d, when it began to look gangrenous. Concentrated nitric acid was applied locally, followed by yeast and lupulin poultices. For the first four days there was constant nausea; and as this was not allayed by the internal administration of opium, the stomach was blistered and dressed with morphia. As soon as the system came under the influence of opium the gangrene was arrested. This was June 1st—eight days; but in that time the whole of the external aspect of the thigh had sloughed to the extent of one foot by eight inches. After the gangrene was arrested, the wound was dressed with tannin gr. v. to glycerine ℥ i., and subsequently with glycerine alone. He had all the stimulants he could take—eggnog, milk punch, beef-tea, &c. He progressed well until June 9th, when he had severe hæmorrhage from one of the perforating branches. This was arrested by ligature of the bleeding vessel. He improved well, and was transferred to New York, June 20th, able to walk on crutches.

Nat. Emory, private, Co. H, 7th New Hampshire Vols., aged 23, received, on May 13th, 1864, a flesh wound of the left thigh from a spent Minié ball. Did well till May 30th, when phagedænic sloughing began, and all the constitutional symptoms of gangrene made their appearance. Ol. terebinth. was used as a local application, followed by yeast and lupulin poultices. Opium being contra-indicated, hyoscyamus was used—two grains of the alcoholic extract every four hours. Tonics and stimulants freely. Gangrene was arrested

June 13th. Slough extended circularly three inches in diameter. Dr. Brower remarks of this case that the wound was slight and the gangrene slight, but the treatment was not sufficiently powerful. Geo. Spitler, corporal, Co. K, 76th Penn. Vols., aged 24, received, on May 11th, 1864, a penetrating flesh wound of the left thigh from a Minié ball. Did well until May 30th, when the wound took on phagedænic sloughing. Concentrated muriatic acid, followed by yeast and lupulin poultices, was used. Opium, tonics and stimulants freely. Dose of opium, one grain every three hours. The gangrene was arrested on June 3d.

The following case of acute mortification, which was also under the care of Dr. Brower, may as well be mentioned in this connection. D. W. Pearson, private, Co. C, 22d South Carolina Reg't, aged 39, received, on June 20th, 1864, a flesh wound of the right tarsus, and also a wound on account of which his right arm was amputated at the middle third on the field. Did well until June 26th, 10, A.M., when the mortification first showed itself in the right foot. The man died at midnight, the entire leg being involved in the destruction. The mortification did *not* attack the stump of the arm. Constitutional condition good. Weight 180 pounds. Always in good health. There was no apparent reason whatever for the mortification. The leg was circumscribed with nitrate of silver, and opiates, tonics and stimulants were given in profusion.

CHAPTER III.

Wounds of the Head and Face.

GUTHRIE has said, with truth, that "injuries of the head affecting the brain are difficult of distinction, doubtful in character, treacherous in their course, and, for the most part, fatal in their results." There is no accident which the surgeon takes charge of with more fear and hesitation; for while in some cases death follows the most trivial injury, in others a vast amount of destruction, and even the removal of brain substance, cause but little disturbance. As Liston says, "no injury of the head is too slight to be despised, or too severe to be despaired of."

The injury inflicted by a ball striking the skull, depends chiefly on the angle at which it strikes, and its velocity. If the angle be very oblique, and the ball somewhat "spent," the injury may be very trivial. Stromeyer asserts that the danger from such a grazing shot arises mostly from pyæmia. Inflammation of the bone following the injury, the veins of the diploe become implicated, and thus pus enters the system. Instances have been related in which the effect of such a grazing shot has been to rupture the vessels between the skull and the dura mater. If the force be greater, the pericranium may be injured to a greater or less extent, in one or other of its tables separately, the fracture of the inner sometimes taking place, while the outer remains to all appearance intact. Lastly, the brain may be injured as well as its case; 1st, by what, as John Bell says, "we choose to express our ignorance of by calling it a concussion," which may pass away doing little or no harm, or may sooner or later result in encephalic inflammation, and compression from effusion; 2d, by such comminution of the skull as shall cause spicula to be thrown into the brain; and, 3d, by direct perforation.

Macleod relates a case, which occurred at the Alma, to show how completely the skull may be destroyed by a glancing shot, without the scalp being implicated. "A round shot, ' en ricochet,' struck the scale from an officer's shoulder, and merely grazed his head as it ascended. Death was instantaneous. The scalp was found to be almost uninjured; but so completely smashed was the skull that its fragments rattled within the scalp, as if loose in a bag."

At a stated meeting of the "United States Army Medical and Surgical Society" of Baltimore, Feb. 19th, 1863, Dr. Geo. W. Dare, Act. Ass't Surgeon U.S.A., related a very remarkable case, where a soldier, having received a perforating wound of the skull, lived for *two months*, the ball remaining in the brain during that time. " Corporal G. W. Stone, Co. A, 12th Mass. Reg't, aged 29, wounded at the battle of Fredericksburg, Dec. 12th, 1862, was admitted into the General Hospital Dec. 19th, 1862. His right eye had been destroyed, as he stated, by a glancing shot. The lower lid was slit to a small extent; the eye was suppurating. He did not complain of much pain. The functions of the other eye were not disturbed, but it was observed to be unnaturally prominent. There were no brain symptoms, except some headache. Neither the patient nor the surgeon suspected that the ball remained in. The wound healed kindly, within the ordinary time. After two or three weeks, the man went out habitually on pass, through the city. He continued well, with the exception of an occasional pain over the remaining eye, until Feb. 6th, 1863, when he was found by Dr. D. in bed, having had a chill followed by fever. The fever assumed a continued form, with occasional chilliness. He remained cheerful, sitting up at times, and did not manifest any serious symptoms until the night of Feb. 10th, when he became delirious. The nurse then reported that he had wandered a little during the two preceding nights, but during the day he had appeared perfectly sensible. Feb. 11th, he was found comatose; the left pupil greatly dilated; intelligence not entirely abolished; he would sometimes answer a question, or put out his tongue when ordered; the vesical sphincter had become involuntarily relaxed. Cupping and blistering were employed without benefit. He sank rapidly, and died about 12 o'clock that night.

"An autopsy was held on the following day. Upon removal of the calvaria, the pia mater was found much congested, and a considerable quantity of dark fluid blood escaped from the torcular herophili. When the anterior cerebral lobes were lifted, a Minié ball was seen lying transversely, half embedded in the sphenoid bone, between its greater and lesser alæ, the concave extremity half way between the crista galli and the sella turcica. A portion of the ball rested against and pressed on the thin inner wall of the left orbit, thus diminishing its capacity, and forcing forwards the eye. The missile had, after passing through the eye, entered the cranium, through the inner wall of the right orbit, at the junction of the sphenoid and ethmoid bones. The ball did not penetrate the dura mater, but remained in contact with, and pressing against it. In consequence of this pressure, ulceration of that membrane resulted, and an abscess formed, extending from the point of contact to the left lateral ventricle, containing several drachms of pus. A few drops of pus, apparently encysted by lymph, were discovered pressing directly the commissure of the optic nerve, which was the probable cause of the extreme dilatation of the pupil. The pons Varolii and the medulla oblongata were found bathed in pus."

I have deferred speaking of wounds of the head causing depression of bone, because I wished, in connection with this subject, to say a few words in regard to the use of the trephine.

In a large proportion of gun-shot wounds of the head, more or less depression of bone is caused. In such a case the question is, what to do. Quesney says, "We should always trephine in wounds of the head caused by firearms, although the skull be not fractured." "All the best practitioners," says Pott, "have always agreed in acknowledging the necessity of perforating the skull, in the case of a severe stroke made on it by gun-shot, upon the appearance of any threatening symptom, even though the bone should not be broken; and very good practice it is." This they would do to relieve compression. But in the first place, we must remember that the brain will often accommodate itself to a considerable amount of depression of bone; and, secondly, that compression caused by the formation of pus consequent to the injury, can rarely if ever be relieved by trephining. The position of the pus cannot be calculated with any degree of accuracy; and few surgeons would care to follow the example of Dupuytren, in his celebrated thrust; fewer still, at the present day, would follow Heister's directions: "Sometimes," he says, "it is impossible to discover the particular part of the cranium which is injured; the patient in the meantime being afflicted with the most urgent and dangerous symptoms. In these cases, it will be necessary to trepan first on the right side, then on the left side of the head, afterwards upon the forehead, and lastly upon the occiput; and so *all around*, until you meet with the seat of the disorder." Moreover, there are numerous instances on record, where men with depressed fractures of the skull from gun-shot wounds have recovered, under circumstances which forbade any attention being paid to them—as on retreats and forced marches. Larrey, Guthrie and Ballingall recount many such instances. And Dease long ago recorded the observation, that "those patients who neglected all precepts, and lived as they pleased, did just as well as those who received the utmost attention." Thus it would appear that all the fatigues and privations of rapid movements in the field are less injurious to men with depressed fracture of the skull, than the probing, picking and trephining of the faculty.

If the bone be very deeply depressed, and the patient comatose, with stertorous breathing, slow pulse and dilated pupil, then it will be advisable to use the elevator cautiously, with or without the aid of Hey's saw. Otherwise, it does not seem to me that any operative interference should be attempted. The operation of trephining is in itself too dangerous to be lightly resorted to—certainly not as a prophylactic.

Injury of the skull, followed at a late date by compression, is, perhaps, the most hopeless of all the circumstances in which the trephine can be used; yet it seems that in which it is most properly employed. As John Bell says, "It is plainly an abscess of the brain; and as it is an abscess which cannot burst and relieve itself, though the trepan may fail to relieve the patient, yet without that help he will infallibly die." It is, in fact, in these cases, *un dernier ressort*.

In the examination of gun-shot wounds of the head, the finger forms the best probe, and even that should be used with much caution. If the ball be lodged in the brain, it should, of course, be removed if possible. Any pieces of bone which are detached and lying on the brain, or which have been driven into it, should be removed with the greatest care, to cause as little disturbance as possible. Cold locally, perfect quiet, low diet, and (if dangerous symptoms make their appearance) purgatives, and bleeding freely repeated, will prove the most useful means of treating such cases. The after-treatment is of the utmost importance. Relapses may occur long after the patient is apparently out of danger, and often from comparatively slight causes. Alcoholic stimuli, retained evacuations, and irregularities of food, are perhaps the most prolific causes.

Wounds of the face are chiefly to be regretted on account of the deformity and disfigurement resulting therefrom. The extreme vascularity of the tissues of the face endows them with a vitality which rectifies most injuries with a rapidity truly marvellous; and from their great distensibility the surgeon is enabled to repair loss of tissue, even when this has been very extensive. The face has been wounded in almost every part and direction, and often presents a most ghastly appearance. The upper and lower jaws, respectively, have frequently been, to a greater or less extent, destroyed, and yet speedy recovery follow. At the battle of Antietam, a soldier had both eyes destroyed by one ball, which passed through the bridge of the nose, leaving a clean hole. He suffered but little pain, and made a rapid recovery.

Hæmorrhage is undoubtedly the greatest source of danger in gunshot wounds of the face; and, from the great difficulty of commanding it, frequently places the patient in imminent danger. Those who have received a severe face wound, seldom leave the field without sustaining a considerable loss of blood; and secondary hæmorrhage is common when the bones have been fractured. The irregularity and extreme vascularity of the parts render the application of ligatures to the bleeding points difficult; and to be effectual, compresses must be applied with much nicety. In secondary hæmorrhage of the deep branches of the face, ligature of the main artery will generally be necessary.

The branches of the facial nerve are sometimes so much injured in face wounds, either by the ball itself or by spicula of bone, that temporary or even permanent paralysis may ensue.

The greatest care should always be taken to remove the secretions which result from injury of the bones of the face. For if any amount of it should be swallowed, and thus enter the stomach, much constitutional disturbance will follow, and a fever of a low typhoid and very fatal type will be induced.

Fractures of the bones of the face form an exception to the general rule, of removing fragments which are nearly detached. The large supply of blood in this region frequently enables pieces of bone —whose direction is not opposed to a proper union—to resume their full connection, in a manner which would be impossible in other parts under the same relative circumstances.

The curious manner in which balls may be concealed in the bones of the face, and be discharged of their own accord, is shown in an instance which occurred at the Alma, and is related by Macleod. "A round ball had entered close to, but below, the inner canthus of the eye, and being lost was not further thought of. The wound healed, and the patient had almost forgotten the circumstance, when, after suffering slightly from dryness in the nostril, the ball fell from his nose, to his great alarm and astonishment, several months afterwards." This case is singular, from the absence of the fœtid discharge which usually attends such injuries of bones, with a retained ball.

CHAPTER IV.

Wounds of the Chest, Abdomen and Bladder.

Wounds of the thorax are very apt to occur in battle, when the combatants are at close quarters. This is to be seen during sieges and street fights. Simple contusion of the walls may be caused by a spent ball, or by a ball which has struck some part of a soldier's accoutrements, or some strong object in the pocket, and has thus been prevented from entering. A case occurred in our army some months since, where an officer was knocked down, but without sustaining any injury; and on examination, he found a rifle-ball embedded in his watch. Such an injury, however, may be so severe as to cause hæmoptysis, severe constitutional shock, and internal inflammation, even though not accompanied by fracture. If the ball strike the edge of any metal plates which form part of a soldier's accoutrements, the part so struck may inflict the injury on the internal organs. Macleod quotes a case from the unpublished records of the English Medical Department, of a soldier "who was hit at Sadoolapore by a round shot on the edge of the breast-plate, which was so turned inwards as to fracture the cartilages of the 5th, 6th and 7th ribs, on the left side, close to the sternum. The skin was not wounded. He walked to the rear, and complained but little for two hours, when he was seized with an acute pain in the region of the heart. His pulse became much accelerated, and he grew faint and collapsed. A distinct and sharp bellows sound accompanied the heart's action. He died in seventy-two (72) hours from the receipt of the injury—the pain and dyspnœa, which had been so urgent at first, having abated for some hours before death. The heart was found to have been ruptured to an extent sufficient to allow of the finger being thrust into the left ventricle. The obliquity of the opening had prevented the blood from escaping into the pericardium, which contained about two (2) ounces of dark-colored serum."

It occasionally happens that a ball is arrested between two ribs. I again avail myself of Dr. Macleod's experience for an illustration. A soldier was wounded by "a large conical ball with a broad base, which was much spent when it struck him. It did not force itself into the cavity, but lay wedged between the cartilages of the 2d and 3d ribs, on the left side, about an inch from the sternum. On withdrawing the ball, the cavity of the chest was found to be fairly opened, and the lung was visible as it expanded and contracted. He had a severe attack of pleurisy a few days afterwards. For five weeks the wound continued to suppurate freely. The lung became adherent to the parietes. The patient made a good recovery."

Fragments of shell not unfrequently open the cavity, but spare the lung; while sometimes the reverse happens, the lung being injured without the pleural sac being opened. It is a singular fact connected with wounds of the thorax, that an intercostal artery is seldom opened.

If the heart or great vessels are wounded, death will almost invariably be instantaneous. A remarkable exception to this rule came under the care of Dr. J. M. Carnochan, of New York. Bill Poole, the noted pugilist of that city, received a wound from a pistol in a fray. Dr. Carnochan was satisfied, on examination, that the ball had entered the heart. The patient, however, did well. About a week or ten days after, he was imprudent enough to see some friends, became excited, and very shortly expired. On *post-mortem* examination, the ball was found embedded in the muscular tissue of the heart.

The dangers which attend a penetrating wound of the lung are, primarily, hæmorrhage and collapse, as well as those arising from suffocation, if the bleeding be profuse. Secondarily, from inflammation and its results, exhausting suppuration and exfoliation, and from the organic diseases that are thereby so apt to be engendered. The collapse which follows these wounds, though dangerous, is the best guarantee for the patient's safety, if not too profound and prolonged.

Balls are well known occasionally to remain embedded in the lung, and become encysted, giving rise to a very slight amount of irritation.

The finger is by far the best probe in examining wounds of the thorax, as in all others, where the wound is not too deep. Fragments of bone and clothing should, of course, be removed with all care; and if the ribs are broken they should be fixed by strips of adhesive plaster, passed from the spine to the sternum, and from above downwards, so as to embrace the wounded side only.

The usual treatment of these cases consists in leaving the wound open, applying light water-dressings, and pursuing, with great vigilance, the antiphlogistic treatment throughout. But in the New York *Medical Times* of Oct. 3d, 1863, pp. 156–7, I find an article by Dr. B. Howard, Ass't Surg. U.S.A., proposing a new treatment, which, if justified by experience, will far surpass the old system. He says : " The custom of leaving the wound open is objectionable, because it affords a means of outflow as fast as the effused blood reaches its level, and thus favors the continuance of *hæmorrhage*. It allows the full force of atmospheric pressure upon the entire surface of the lungs, and thus necessitates *dyspnœa*. It admits continually renewed currents of atmospheric air, ensuring decomposition of the clot in the pleural cavity, with extensive and profuse *suppuration*, of a very fœtid character, while it does not provide for its exit, until after so great an amount has accumulated as to cause it to rise above the level of the wound; and after its partial subsidence by overflow, the wound again ceases to be available.

" Suppose, however, that the wound be perfectly closed, the following will at once appear among the advantages to be gained. 1st, *hæmorrhage* is controlled. At the worst, the amount of blood lost after the operation cannot be more than would suffice to fill up the unoccupied space in the pleural cavity; the elastic clot resulting, furnishing a styptic, par-excellence, for the wounded vessels of the yielding lung. 2d, dyspnœa is immediately relieved, upon removal of the atmospheric pressure, and the restoration of the parts approximately to their normal condition. The enclosed volume of air being absorbed, the lung is again at liberty to expand with its usual freedom, limited only in proportion to the size of the clot, which may happen to be in the pleural cavity. 3d, *Suppuration*, if not prevented, is greatly diminished by shutting out the constantly renewed currents of atmospheric air, and its character is very favorably modified. Indeed, if the wound were closed soon enough, I deem it possible that the slough of the track through the lung, with the limited amount of attendant pus, might be entirely disposed of by absorption and expectoration. The operation which I practise, is by hermetically sealing, as follows :—All accessible foreign bodies having been removed, introduce the point of a sharp-pointed bistoury, perpendicularly to the surface, just beyond the contused portion, and with a sawing motion pare the entire circumference of the wound, converting it into a simple incised wound, of an elliptical form; dissect away all the injured parts down to the ribs; then bring the edges of the wound together with silver sutures, deeply inserted, and not more than a quarter of an inch apart; secure them by twisting the ends, which are then cut off short, and turned down out of the way. Carefully dry the surface, and with a camel's hair pencil, apply a free coating of collodion over the wound; let it dry, and repeat it at discretion. For greater security, shreds of charpie may now be arranged crosswise over the wound, after the manner of warp and woof; saturate it with collodion, and when dry, repeat the process until the wound is securely cemented over; as a still greater security, a dossil of lint may then be placed over the part, and retained with adhesive straps.

" If there be a tendency to undue heat in the part, it may be kept down with cold affusion; should any loosening of the dressing occur, an additional coating of collodion may be applied. The sutures must not be removed, until healing by *first intention* is complete. Should suppuration occur, so as to occasion distressing dyspnœa, proceed to treat it, in all respects, as a case of empyema, introducing the trochar at the most dependent point, and taking especial care to avoid the admission of air. . . . . . . In incised or punctured wounds, the paring process is, of course, dispensed with. . . . . . .

" Practically, the immediate results have been very remarkable, and, I think, unprecedented. The most painful cases of dyspnœa have been promptly relieved, the patient usually falling into a quiet slumber, in about an hour after the operation. . . . . . . I have obtained healing by first intention, and removed the sutures, within five (5) days after the operation. . . . . . . Some cases upon which I operated, were six days in the ambulances, before reaching a General Hospital, part of the road travelled being of the worst description; on the fifth day, all but one of these, so treated, were able to walk comfortably. On their arrival, all the wounds were unfortunately reopened, except where the union was too complete to allow of it, and the usual water dressing was substituted."

This treatment applies to penetrating wounds of the abdomen, as well as to those of the thorax. In fact, Dr. Howard's first experiment was in a case of bayonet wound of the abdomen, "which was followed by the best results."*

The abdominal cavity, from its large surface, and want of bony protection, is very liable to receive injury in battle. And wounds of this part, are of the most dangerous character.

Contusions by round shot or shell, are among the most severe injuries to which the abdomen is exposed. The solid or hollow viscera may be ruptured thereby, and rapid death follow, with but little external appearance of so grave an accident. And even if not so severe as this, injuries of this sort are not uncommonly followed by extensive sloughing of the abdominal wall. If the amount of inflammation caused by contusion be limited, adhesion will be likely to take place between the parietes and the omentum, or viscera, and thus afford a great safeguard against the effusion of blood or other matters into the peritoneal cavity.

Vomiting and pain in the abdomen are the usual symptoms which contusions of this cavity give rise to. If any internal rupture has taken place, we can do very little to prevent a fatal issue. Otherwise, the treatment should be such as will ward off peritoneal inflammation, which may steal on very insidiously.

Penetrating wounds of the liver, kidneys or spleen are, as a rule, fatal; though many exceptions occur. Wounds of the stomach are also exceedingly fatal. Baudens (Observation iv. p. 12, of his "Clinique") records a remarkable case of recovery, though complicated with severe head injuries. Wounds of the small intestines are said to be much more serious, than those which injure the large.

The wonderful manner in which balls and swords may traverse the abdominal cavity, without wounding any of the viscera, has often been commented upon by military surgeons.

When a ball merely enters an intestine and lodges there, it may sometimes be thrown out by the rectum. Several such cases have been related in the New York *Medical Times,* during the present war.

If a vascular viscus or large artery have been wounded, hæmorrhage to a very serious, and even fatal extent, may take place within the cavity. The mutual pressure of the viscera, however, does much to prevent the former accident; and the slight attachment of the arteries generally enables them to escape.

Early protrusion of the intestine is rare, unless the wound be quite large. It should, of course, be immediately and carefully returned, when it does occur.†

---

* Since the above was written, I have been told by Army Surgeons who had tried the method proposed by Dr. Howard, that they had completely failed in obtaining the good results claimed by him.

† The treatment of such cases can only consist in absolute quiet—with the exhibition of opium, if necessary—and in sustaining the powers of life to the utmost. But they are very hopeless cases to treat.

The principal danger attending penetrating wounds of the bladder, arises from the infiltration of urine, either primarily or secondarily. If the bladder be full when struck, rapid death will be almost inevitable, on account of the infiltration of urine into the peritoneum. If, however, the bladder be empty, it becomes a matter of great importance to introduce a gum-elastic catheter, as early as possible, and let it remain, so that no accumulation of urine will be possible. The catheter should be allowed to remain until we have reason to believe that the wound is *thoroughly* closed. It presence will excite but little irritation, unless the neck of the bladder be injured.

If the ball have remained in the bladder, it becomes a matter of consequence to remove it. Balls and pieces of cloth or bone, so retained, have in many instances become the nuclei of calculi; so that the sooner they are removed the better. Demarquey mentions a case in which the nucleus was a piece of shell. Larrey operated, successfully, four days after the introduction of the ball into the bladder. In some cases, where the ball was small, it has been voided with the urine.

A case from one of the battles of the Peninsula, in the year 1862, came under my care (of which, unfortunately, I have lost the notes), where a soldier while stooping over, with his back to the enemy, received a ball about three fourths of an inch to the left of the anus. Upon examination, I found the ball in the bladder, and by merely enlarging the track of the ball, performed the lateral operation of lithotomy. He did excellently well, and was removed to one of the northern hospitals in about three weeks.

CHAPTER V.

Compound Fractures of the Extremities, Amputations, &c.

Gun-shot fractures of the femur are of very frequent occurrence in battle, and are remarkably dangerous, as compared to fractures, not compound, of the same bone, such as are met with in civil practice. " All the complete fractures of the other bones of the extremities unite, when well managed; by what fatality are those of the femur not equally fortunate ? Is it the diameter of the cavity of the bone; the quantity of medullary substance which it contains; the peculiar structure of the vessels which carry the nourishment; the size and force of the muscles which are attached to it, which by their weight and pressure obstruct the passage of the liquids ? All these causes united, may combine together and give rise to that want of success which we meet with in treating complete fractures of the femur caused by firearms; but complete fractures of this bone heal very well, whatever cause has produced them, when they are not accompanied by a wound."—(Ravanton, *Chir. d'Armée*, p. 324.)

4

All the dangerous characteristics of gun-shot fractures have been greatly increased by the conical ball. 1st. The shock it occasions is far greater than that caused by the round ball, merely because the destruction resulting therefrom is much more severe. 2d. The comminution of bone is enormously increased. 3d. The bruising of the soft parts is more extensive; consequently the suppuration is more prolonged, and the chances of purulent absorption so much increased.

There has been a great cry throughout the profession, of late years, for conservative surgery; and this is just and right, and has undoubtedly done much good. But there are many concomitant circumstances in the military practice of surgery which force us to amputate, when in civil practice we might save the limb. And this is especially true in gun-shot wounds of the femur. In the first place, the standard of health and strength is much reduced in those who have for any length of time been subjected to the privations and hardships of campaign life; they live so completely up to their income of health—so to speak—that when any additional drain is made on the system, by a wound or sickness, their strength fails very rapidly. And, in the second place, it is essential to the successful treatment of compound fractures that the patient be supplied with suitable food; that his broken limb be retained, for a certain length of time, immovably fixed in a proper apparatus, and that careful treatment should be continued for a length of time. But how can these things be accomplished in war? Field-hospitals are not overflowing with comforts and conveniences, and the transportation to a general Hospital is often long and tedious—never more so than in the present war. "Thus we foresee," says John Bell, "an argument of necessity, as well as of choice, and that limbs which in happier circumstances might have been preserved, must often in a flying army or dangerous campaign be cut off. It is less dreadful to be dragged along with a neat amputated stump, than with a swollen and fractured limb where the arteries are in constant danger from the splintered bones; and where by the least rude touch of a splinter against some great artery the patient may in a moment lose his life." And Dupuytren says, in one of his clinical lessons, "I have repeated it often, and I repeat it for the last time, after the facts which I have observed, chiefly in 1814, 1815, and 1830, that my opinion on this point is unshaken. In compound fractures from gun-shot, in rejecting amputation, *we lose more lives than we save limbs.*"

According to Macleod, the lower extremity was removed at the *hip* ten times during the Crimean War, in the English army, all primary operations, and all ending rapidly in death. The same author says, "Although making every endeavor, I have only been able to find a record of three cases in which recovery followed a compound fracture in the upper third of the femur, without amputation. . . . . I am certain, however, that although the instances of recoveries were rare, they were yet not so exceptional as recoveries after amputation of the same part. . . . . After the 1st April, 1855, amputation in the upper third of the thigh was performed 39 times, with a fatal result in 34 cases. Of the total number only one was a secondary operation, and it ended fatally. . . . . Amputation in the middle third was performed, during the period after the 1st April, 1855, 65 times, of which 38 died; 56 of these cases, and 31 deaths, were primary operations; 9 cases were operated on at a later period, and 7 died. Amputation was performed on the lower third, during the same period, 60 times, 46 being primary and 14 secondary operations; of the primary, 23 died; and of the secondary, 10." Thus we have 174 amputations in all parts of the thigh, and 115 deaths. It would appear, then, that in gun-shot fractures of the upper third of the femur we should try to *save* the limb; but in a like injury of the middle and lower thirds, primary amputation will be our best course.

As a general rule, I think the circular operation will be found the best in amputations of the thigh; 1st, because by this operation we can remove the limb further from the trunk than by the flap operation; and, 2d, because in transportation we escape the knocking about and consequent loosening of the heavy flaps, which so frequently occur after the last-named method.

Gun-shot wounds of the leg, fracturing both bones, will generally require amputation; and the sooner it is performed the better. The statistics of the Crimean War, after April 1st, 1855, give 101 amputations of the leg, and 36 deaths resulting therefrom; 89 cases and 28 deaths were primary operations; and 12 cases and 8 deaths, secondary.—(Macleod.)

Where one bone only is fractured, the limb should be saved, even if that one be considerably destroyed, as the other will serve to steady it during the repairing process. When the fracture occurs near the ankle the injury is more severe than when it takes place near the middle of the bone. Resections may sometimes be performed on one or other of the bones of the leg to advantage.

Compound fractures of the upper extremity are not, as a rule, so dangerous as those of the lower, and much greater success will be met with in their treatment. Several causes combine to produce this result, among which the following may be named—the free anastomosis which exists between the vessels, the large supply of blood which they convey, the smaller amount of suppuration, and the less necessity for the patient to keep in a constrained position during recovery.

In the Crimea, after the 1st April, 1855, amputation at the shoulder was performed 39 times, with 13 deaths: 33 cases and 9 deaths were primary operations, and 6 cases and 4 deaths were secondary. Amputation of the upper arm was performed, during the same period,

102 times, with 25 deaths; 96 cases and 22 deaths were primary, and 6 cases and 3 deaths were secondary. The forearm was amputated, during the same time, 59 times, with three deaths; 52 cases and 1 death were primary, and 7 cases and 2 deaths secondary.—(Macleod.) Thus we have 200 amputations of the upper extremity, with 41 deaths.

Resections are particularly applicable to the upper extremity, the arm retaining its full power of motion in many ways, and increasing in power and usefulness with practice.

Wounds of the foot and hand, even of the most severe character, frequently make very good recoveries; and this result should always be aimed at. Even if partial amputation be necessary, we can often contrive to leave the patient a very useful remnant.

The special methods of amputation, and of the treatment of compound fractures, when we try to save a limb, do not require any particular mention in an essay of this size and character.

<h2 style="text-align:center">CHAPTER VI.</h2>

Wounds of Joints; Excision of Joints, &c.

The gravity of gun-shot wounds of the joints depends chiefly on the size and construction of the joint, the extent of the injury, and the conveniences for careful treatment which may be at hand. The wound of a ginglymoid articulation is generally more severe in its consequences than that of a ball-and-socket joint—principally on account of its more complicated structure. Larrey makes mention of the frequency of tetanus as a consequence of wounds of these joints.

Even when there has been a very great amount of destruction of the articulating extremities of bones, the external wound and appearance may be so trivial as to deceive us in regard to the nature of the wound, and often induce us to delay the prompt and decided measures which are so necessary to ensure recovery.

The hip-joint is so deeply placed, and so much protected by the surrounding parts and its own form, that it is not often penetrated by a ball; but when this does occur the destruction is usually very serious. Alcock lost three out of four cases in which this accident occurred; and in the fourth case, " where recovery took place, the joint itself, there is some reason to suspect, was but remotely affected."

A round ball may occasionally become impacted in the head of the femur, with or without a partial fracture of its neck. It is very difficult, however, to recognize such injuries at first, as there may be neither displacement nor crepitation perceptible. Larrey mentions the case of an officer, wounded in Egypt, who received a ball in the neck of the femur. The wound closed, and twenty years afterwards, on the death of the patient from disease of the chest, the ball was found impacted in the bone.

A penetrating wound of the knee presents an injury of the very gravest description. Macleod says, in referring to wounds of this joint, "I can aver that I have never met with one instance of recovery in which the joint was distinctly opened, and the bones forming it much injured by a ball, unless the limb was removed." A round ball will sometimes penetrate the lower end of the femur, or head of the tibia, without opening the joint, or at least with very slight injury to the capsule, and these cases may recover; moreover, balls may pass close to the capsule, and yet do it no harm, and it is such cases as these which are frequently recorded as recoveries from penetrating wounds of the joint. It is, undoubtedly, often very difficult to determine whether the joint be opened or not, particularly if the ball be a small one. It often happens that the round ball will course around the bones superficially, when it appears to have passed directly through the articulation. We should remember, also, that the swelling of the joint may be caused merely by a bruise, or by the extension of inflammation from some neighboring injury.

When the bones are not much injured a considerable length of time may intervene before severe symptoms set in; and this again may tend to deceive us in regard to the real nature of the injury.

Joint wounds from conical balls, however, are not apt to leave us in much doubt. They generally crash into and through whatever comes in their way, leaving very distinct marks of their progress.

The danger of these wounds is not immediate, but lies in the long wasting suppuration, the dangerous abscesses which burrow far into the neighboring tissues, and the manifold chances of pyæmia. These abscesses appear among the muscles of the thigh, frequently burrowing along the bone, and stripping it of its covering, and yet are seldom—so far as we can see—in connection with the joints. They are often unnoticed for a long time, and give rise to a great deal of trouble and danger. At a late period of the case, the joint puts on all the appearance of white swelling.

Military surgeons have always acknowledged the necessity of early amputation, where the articulating ends of bones have been fractured by a ball. Esmarch, from the field of Schleswig-Holstein, says, "All gun-shot injuries of the knee-joint, in which the epiphysis of the femur or tibia has been affected, demand immediate amputation of the thigh. It is a rule of deplorable necessity, already given by the best authorities, and which our experience fully confirms." Guthrie has seen no case recover in which the limb was not removed.

Macleod says: "I have often contemplated the laying of the articulation freely open, at an early period in these cases, so as to permit of the extraction of all foreign bodies, and the free escape of the pus which must afterwards be formed, the retention of which is undoubtedly one great source of danger. This might be attempted, even although it were necessary to lay the whole front of the joint open, by an incision similar to that for excision. The joint has been frequently widely laid open by cutting instruments, both primarily, and for disease, and most satisfactory results have been obtained. [*Note.*—At the time the above was written, I had not seen Stromeyer's book, and did not know that the same idea had occurred to him, or that in the only case in which he had practised it, the results had been most encouraging.] "

In July, 1864, while at Hampton Hospital, Va., a case came under my care in which there had been an apparently slight shell wound of the outer side of the left knee. The wound was small and the joint not opened. About two weeks after the receipt of the wound, however, it was found that pus had burrowed into the joint. My own strong desire was to amputate immediately; but upon calling a consultation I was over-ruled, and it was determined—the patient being young and very healthy and robust—to try to save the limb according to Stromeyer's and Macleod's suggestion of laying open the joint. This was done by a free incision on each side of the patella—the limb hung in Smith's anterior splint, and everything done to improve his general condition. The patient did very well until about the eighth day; he then sank very rapidly, and died on the tenth day after the joint was opened.

No better or fairer case could be selected upon which to try this plan of treatment; but it failed most lamentably, and only served to give more force to the rule that where the knee-joint is opened by a gun-shot wound, either directly or by ulceration, the limb should be *immediately* amputated. If you wait and hope, your patient does well for a little while, and then sinks with such fearful rapidity that no power or science on earth can save him.

If we decide to make the attempt to save the limb, the most rigid antiphlogistic treatment should be enforced; as well as the early evacuation, by free incision, of abscesses, and of matter if it form within the joint. Hectic, with its common accompaniment—diarrhœa—pyæmia and tetanus, are the causes which generally bring death to the relief of the sufferer.

If amputation be decided on, the sooner it is performed the better; for if the operation be deferred, until inflammation and suppuration have been for some time present, the results are very unfavorable.

Shell wounds of the knee are not usually so dangerous as bullet wounds. They frequently merely cut the soft parts open; or, if they injure the bone, they leave a larger aperture, by which the secretions can be more freely discharged. These wounds frequently recover well, with more or less anchylosis.

Macleod has seen only one gun-shot fracture of the patella. The bone was " starred," but the ball did not lodge. The recovery was good, " the motion of the joint being, however, considerably interfered with."

Penetrating wounds of the ankle will generally do well if properly treated, but they require much and long-continued attention. Two things are especially required: 1st, that the articulation be rendered perfectly immovable; and, 2d, that one or other of the wounds be so enlarged as to allow of the free escape of all discharges.

The shoulder-joint, from its more simple mechanism, will, as a rule, suffer less, and from its superficial position will be more manageable when injured than almost any other articulation. Balls sometimes pass very close to the capsule without opening it, or with very slight injury. Larrey tells us that he saved many cases, in which the opening into the joint was not great. If a ball remains impacted in the head of the bone, as sometimes happens with the round ball, it should be removed as soon as possible; as, otherwise, caries of the bone and disease of the joint will be the probable result. Abscesses and fistulous tracks are the concomitants most to be dreaded in penetrating wounds of the shoulder-joint.

Wounds of the elbow are more dangerous than those of the shoulder, on account of the more complex nature of the joint. Excision will be found the best and most satisfactory method of treatment.

We now come to speak of *Excision of Joints*—a subject very interesting to the military surgeon, on account of the great number of cases in which it can, with propriety, be substituted for amputation; and even, in some instances, holding out a far better hope than the last-named operation.

In the Crimean war, after the 1st April, 1855, excision was performed on

Head of the femur . .	5, primary, of which 1 recovered.	
" " " . .	1, secondary, fatal.	
Knee-joint . . .	1, " "	
Os calcis and part of astragalus	1, recovered.	
Os calcis alone . .	1, "	
Head of humerus . .	8, primary, 1 death.	
" " " . .	5, secondary, no death.	
Do. and part of scapula .	1, " fatal.	
Elbow-joint . . .	13, primary, 3 deaths.	
" " partial . .	3, no death. (Macleod.)	

The same author says: "The shafts of the bones leading from the joints were often too extensively destroyed to enable the injured parts to be removed by excision; in fact, the shafts were so often split, and their periosteal and medullary membranes destroyed, that the resection of the articulation did not suffice to save the limb. Surgeons soon recognized this; but yet it was by no means always easy to determine the true state of things about the joint till the incisions necessary for resection laid bare the bones, and forced the reluctant operator to convert his operation into one of amputation."

Primary excisions will be found much more successful than the secondary, both as regards the final results, and the length of the period of convalescence. Very much depends on the after-treatment; especially in guarding against inflammation.

The shoulder-joint is certainly that to which excision is the most applicable; both from its simple construction, its superficial position, and from the greater readiness (on account of the great vascularity of the surrounding tissues) with which the reparative process is here accomplished.

Guthrie thought the insertion of the deltoid the lowest point at which the bone could be divided, with any prospect of success. But Esmarch has shown, that as much as four and a half inches may be removed from the humerus, and yet a very useful arm remain. Of this fact I know several instances resulting from the present war. Moreover, the report of Esmarch on the practice of Stromeyer and Franke shows us, that to cut across the fibres of the deltoid does not much interfere with its after usefulness; "as its upper edge applied itself to and united with the articular surface of the scapula, and was thus fully attached, and able to raise the arm. The healing was also quicker, as the space to be filled by granulation was much diminished in size, by the application of the muscle to the glenoid fossa."

The fact so clearly demonstrated by Stromeyer, should always be borne in mind when determining on operations at the shoulder-joint; that in comminution of the shaft of a long bone, the fissures never extend into the epiphysis; in the same manner, injuries of the epiphysis, only in very rare cases, extend into the shaft, unless the bullet strikes the adjoining borders of both parts.

It will be found, that in the field we do not require such extensive incisions as in civil practice; and thus, of course, the hope of restored action will be much increased.

One of the chief dangers resulting from excision of the shoulder, lies in the formation of abscesses and sinuses in the neighborhood. To avoid this, we should arrange the line of incision so as to give free exit to the pus; and with this object in view, we shall find Stromeyer's semicircular incision over the posterior surface of the articulation better than any other.

Resection of the elbow has long been known to be far less fatal than amputation. The question now is, how much of the articulating ends of the bones can be removed, consistently with retaining a useful joint? Esmarch thinks that "the less there is removed from the joint ends of the bones, the greater is the probability of anchylosis." But there is considerable diversity of opinion, as to the greatest amount of bone which should be removed.

The complete fixture of the joint during the earlier period of treatment, which is so emphatically dwelt on by Stromeyer; its constant support by a splint, even when being dressed; the elevation of it, to prevent œdema, and its flexure at an angle of 130° to 140°, are all points of great importance. Before the wound has wholly cicatrized, passive motion should be attempted; but it must be at once abandoned, if any irritation or sign of inflammation makes its appearance.

Macleod tells us, that so far as he knows, the knee was only excised once in the Crimea, and that patient died from causes unconnected with his wound. Mr. Ferguson thus sums up the advantages which his large experience ascribes to the operation:—" The wound is less than in amputation of the thigh, the bleeding seldom requires more than one or two ligatures, the loss of substance is less, and probably, on that account, there is less shock to the system; the chances of secondary hæmorrhage are scarcely worth notice, as the main artery is left untouched; there is, in short, nothing in the after-consequences more likely to endanger the patient's safety, than after amputation, whilst the prospect of retaining a useful and substantial limb should encourage both patient and surgeon to this practice."

Resection of the hip, though rather a hopeless operation, is still far more hopeful than exarticulation. Thus we see from the returns of the Crimean war—of ten cases of amputation, all died; while in six cases of excision, one recovered. The great danger of these operations lies in the after-treatment. As in all other excisions of joints, it is very necessary to keep the joint immovable during the early period of treatment, to provide free exit for the pus, which is sure to make its appearance from abscesses and sinuses in the muscles of the thigh, and to guard against inflammation.

I am fully aware that field-hospitals do not contain all the conveniences for treating these cases; but in very many cases, excision gives us a better chance of saving life, than we can hope for from amputation, besides rendering the life, thus saved, very much more comfortable to the patient. And that the life and permanent welfare and comfort of our patients is the principal object of our profession, no one will deny.

www.ingramcontent.com/pod-product-compliance
Lightning Source LLC
Chambersburg PA
CBHW022033190326
41519CB00010B/1698